Institute of Terrestrial Ecology
NATURAL ENVIRONMENT RESEARCH COUNCIL

Behaviour of Badgers

Hans Kruuk
and
Timothy Parish

The Institute of Terrestrial Ecology (ITE) was established in 1973,
from the former Nature Conservancy's research stations and staff,
joined later by the Institute of Tree Biology and the Culture Centre of
Algae and Protozoa. ITE contributes to and draws upon the collective
knowledge of fourteen sister institutes which make up the *Natural
Environment Research Council,* spanning all the environmental
sciences.

The Institute studies the factors determining the structure, composition
and processes of land and freshwater systems, and of individual plant
and animal species. It is developing a sounder scientific basis for
predicting and modelling environmental trends arising from natural or
man-made change. The results of this research are available to those
responsible for the protection, management and wise use of our
natural resources.

Nearly half of ITE's work is research commissioned by customers,
such as the Nature Conservancy Council who require information for
wildlife conservation, the Forestry Commission and the Department
of the Environment. The remainder is fundamental research supported
by NERC.

ITE's expertise is widely used by international organisations in overseas
projects and programmes of research.

Published by Institute of Terrestrial Ecology
68 Hills Road, Cambridge CB2 1LA
0223 (Cambridge) 69745

ISBN 0 904282 11 2

Badger problems in Britain

The badger is the largest carnivore in Britain and also one of the commonest. It is almost the symbol of nature conservation in the country, indeed a special law has been passed for its protection. Nevertheless, very little is known about its ecology, about its environmental requirements, about the factors that regulate its populations, and about its social behaviour. Such knowledge is essential, however, if we are to manage this species in problem situations, or if we are to protect it in nature reserves.

The badger's relationship with man is ambiguous. On the one hand, it is one of few carnivores which have benefitted from environmental changes brought about by man.

Badger ♂.

Throughout the country it is often found in close association with old agricultural development, and it feeds extensively on pasture land which provides it with an unusually rich source of food – earthworms. On the other hand, man is the only serious enemy of badgers in this country, persecuting the animal for the sake of its skin, for sport ('badger baiting'), or to prevent its real, or alleged, interference with livestock and game birds. The most serious cause for concern in our relations with badgers is their ability to transmit important diseases, such as bovine tuberculosis and rabies.

In order to take effective action when badgers are involved

1

as vectors of disease, it is essential to have a basic knowledge of the animals' ranges, of their social contacts, and of their responses to management. Whether for the purpose of conservation or of control, a basic knowledge of the badger's ecology and behaviour is required, and at present we do not possess this information.

The present research project

The badger project which was started in April 1975 at the Institute of Terrestrial Ecology laboratories at Banchory aims at filling some of the gaps mentioned above, carrying out basic research to provide the information that can assist management of badger populations in problem situations and for conservation purposes. In doing so, we hope to shed some light also on more theoretical issues, on the effects of the environment on animal behaviour, and on the evolution of social organisations. The project provides an ideal opportunity to understand biological issues of fundamental importance, whilst at the same time providing results which can be immediately useful on a more practical level. Furthermore, the team involved in this study is developing expertise and techniques within the Institute which are clearly essential in studies of other carnivores and related types of field work. This is an important aspect, since problems with carnivores will be with us in Britain and elsewhere for a long time to come.

Habitat of one of the Scottish study areas: Ardnish.

The main questions which are asked in this study are designed to get an understanding of the factors that determine badger numbers. Our interests could be expressed thus:—

(a) How are badgers organised?
     ie What determines group size: group composition (sex, age, dominance): reproduction and mortality: territory size, overlap, defence?

(b) How do badgers use the environment?
     ie What determines food selection (diet, available food, feeding behaviour): selection of sett sites?

What is the effect of (b) on (a), and vice versa?

This account briefly describes some of the work we are doing to answer these questions, and some of the results achieved so far. Basically, we try to compare badgers in a variety of study areas, which have been selected for their different habitats, or their gross differences in badger densities; we

Habitat of one of the Scottish study areas: New Deer.

also look at details of the animal's behaviour in captivity where we can manipulate some of the variables in which we are interested.

How is a badger population organised?

Badgers live in groups of variable size, which we have termed 'clans'. Each clan has as its headquarters a main sett with several entrances, and is usually occupied continuously; apart from this main sett, single outlying holes are occupied irregularly. In our study areas, we have looked first at the distribution of the main setts, especially how far they are apart. Figure 1 shows the distances of the nearest neighbours to every known main sett in some areas. These provisional results show that the distances vary in the different study areas and from this kind of information we can establish the number of setts per unit area.

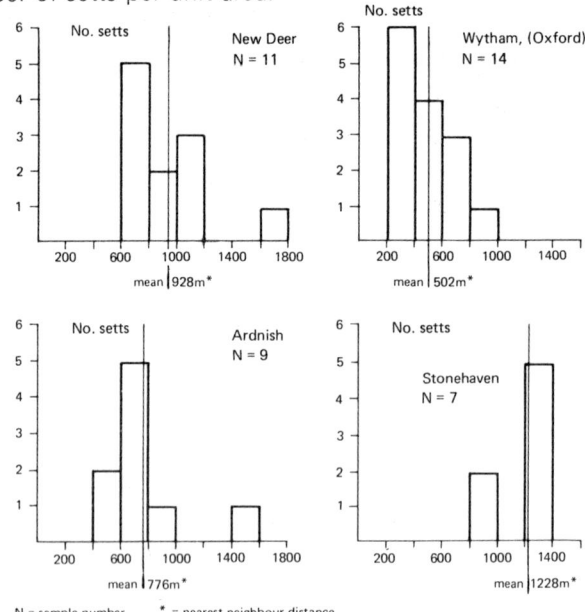

N = sample number   * = nearest neighbour distance

Figure 1. Distances between setts in some of the study areas. The histograms show numbers of setts which have the nearest neighbours at a certain distance.

Apart from the differences in distances between setts, the numbers of badgers per clan may also vary, and this information is needed to estimate the population density in a given area. Just counting badgers emerging from their setts at dusk is likely to provide a considerable under-estimate and to overcome the various drawbacks to this method, we are developing a digital recording system which can be used for counting animals in setts with many entrances, over periods of several days.

The next questions to ask are about the badger's ranging behaviour. How far do they move from their setts, how do animals from different setts tolerate each other in an area, which vegetation types do they use and how? To answer these and other questions, it is necessary to be able to follow badgers on their wanderings at night, which is possible with the aid of radio-location. In this technique, a small radio-transmitter is attached to the badger, on its shoulders, kept in place by straps of leather around neck and chest. The transmitter produces 40–70 beeps per minute, which are audible on our portable receiver at distances of up to several miles, and with a directional aerial we are then able to locate the animal. Attached to the transmitter is a small 'beta-light', which emits a greenish light comparable to the light of a few glow worms together; this makes it easier to spot the badger in the dark once the radio has enabled us to come

The end of a badger: one of the radio-collared animals, killed in a game-keeper's snare.

Badger with radio-transmitter on its back.

close enough. Although the radio-tracking system is a great help, it is still difficult to follow the badgers around at night; it takes a considerable amount of practice and experience to observe the animals when they are away from their setts, and the observer has to take great care to stay down-wind and to move unobtrusively.

One of the results from the radio work is a knowledge of the range of badgers and the holes in which they spend the day. In Wytham Woods, near Oxford, we found that badgers occupied ranges of 50–150 hectares, ranges which usually did not overlap with those of the neighbours, but which were the same as those of other badgers from the same clan. In some clans, however, ranges of males overlapped with those of females from two different setts. The ranges of the females from the same sett overlapped completely, but excluded the range of the females from the other sett. Apart from this, there was a group of bachelor males without females, who also defended a communal range, but this range was less than half the size of a normal clan. Indications are, however, that this system is by no means universal in badger populations, and we are likely to find a very different social organisation in

Badger in trap. The animals are caught for the purpose of putting on radio-transmitters.

the less densely populated areas of Scotland. Already we have found one old male, presumably a 'bachelor' ranging over an area of more than 1,200 hectares, including several main setts, something that would be unexpected in the high-density population near Oxford.

Importance of social behaviour

The signals from the radio-transmitters can be picked up also when the badgers are asleep underground, during the day-time. From checks on where the various badgers spent their days, we know that, especially in the summer, the females and young males often sleep in the single entrance outliers, rather than in the main setts, while the large dominant boars could always be found in the main sett. It is from this kind of evidence that we assume that there is a social structure in the badger community which might be related to population limitation; some animals appear to be 'on the social fringe', which might affect their reproduction and survival.

The radio transmitters are used not only for merely establishing the badgers' ranges, but also to find animals in

In the sleeping quarters of the badger enclosure at Hill of Brathens research station the animals can be watched underground as well as in semi-natural vegetation.

the field, in order to observe their social and feeding behaviour. The badgers are watched through infra-red night glasses (the 'hot-eye'), and we are now beginning to understand something of the way in which the badgers' ranges are maintained. The species is highly territorial, at least in the high density areas of southern England, marking its range boundaries with conspicuous 'latrines' and defending these borders with aggressive behaviour. How this works in less densely populated areas is still unknown, but clearly territorial behaviour plays an important role in the distribution of badgers throughout their range.

Despite our use of equipment which makes observation of badgers in the wild possible, it is still very difficult to establish exactly what is happening within a badger clan. This is because for most of the time badgers are out of sight, either in the vegetation or underground. To complement the restricted possibilities of observation in the wild, we are keeping a group of captive badgers at the Hill of Brathens Research Station. Here we can watch the animals in a large enclosure with natural vegetation, and we can observe them inside their sett. We are trying to discover what kind of behavioural organisations exist between our animals, how this

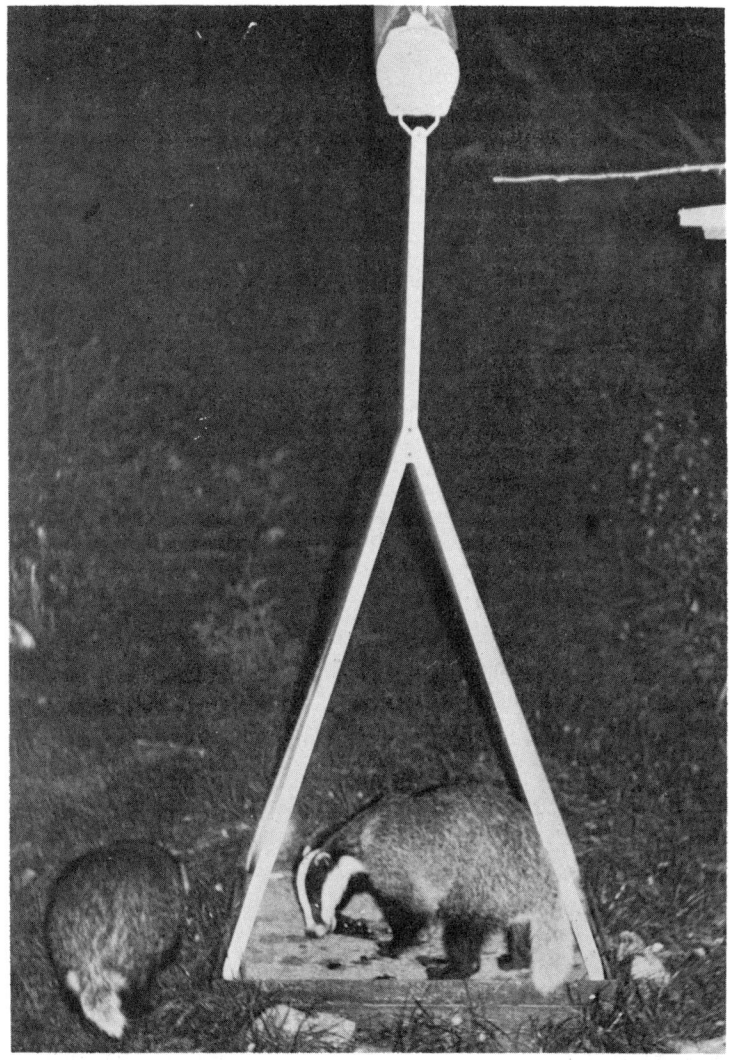

In the enclosure for captive badgers at Hill of Brathens research station: animals are weighed regularly, to study the relation between weight and social status.

affects their reproduction, how strangers are treated, and exactly how badgers communicate and recognise each other. Are the scent secretions of one clan different from those of the neighbours? If so, would we be able to identify these differences from samples collected in the field? It is especially for the purpose of solving these questions of

9

Food: the earthworm (Lumbricus terrestris), the most important food of the badger, feeding on the surface at night (tail in burrow).

Food: wasp nest, immediately after it was uncovered by a badger.

Food: ears of wheat, after badgers have finished with them.

scent-communication, which appear to be very important in the life of the badgers, that ITE is collaborating with the Department of Zoology in Aberdeen University; the secretions of various glands of the badger are being analysed there, whilst at the Hill of Brathens Research Station we look at the behavioural implications of scent-gland activities.

Food and foraging

Almost invariably, in the past, badgers have been described as 'omnivores' eating things as widely different as wasps and

11

The analyses of badger-faeces in the laboratory.

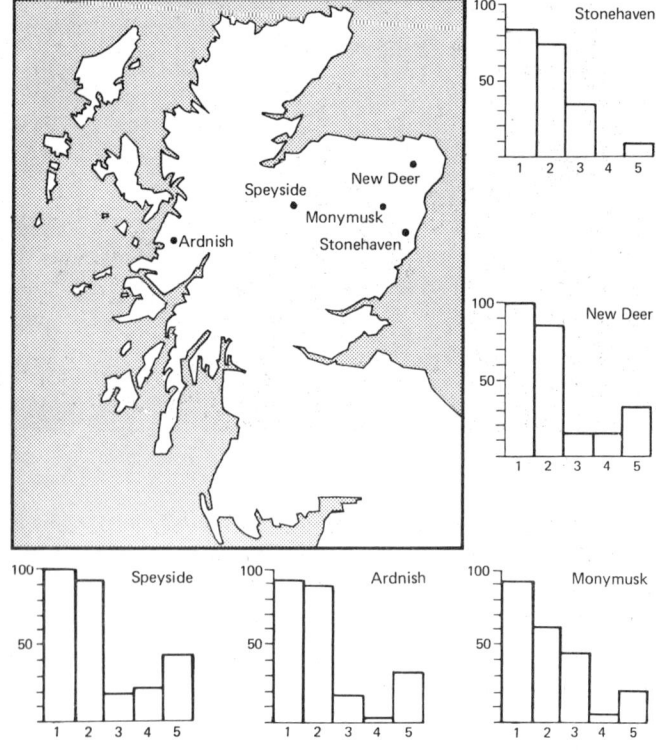

Figure 2. Badger food in Scottish study areas. The histograms show percentages of badger faeces in which various food items occurred. Key: 1 = earthworms, 2 = beetles, 3 = rabbits, 4 = oats, 5 = pig-nuts.

wheat, worms and hedgehogs, carrion and fruit. Although this appears to be true also in the badgers which we have studied, it is nevertheless a distorted truth; the European badger is, in fact, a highly specialised feeder who does eat all sorts of things, but only 'on the side'.

One way to collect information about an animal's diet is through analysis of its faeces. We have collected many hundreds of samples from our different study areas; these are taken to the laboratory, where they are washed and the contents are identified and measured. Some of the results are shown in Figure 2. They are a clear indication of the differences in importance of the various kinds of food in the different study areas. However, even more striking is the *similarity* between the badger diets of different places; in all the habitats, earthworms are by far the most important food, which is especially clear when one takes into account the quantities with which each of the different kinds of prey is represented in the various droppings. Rabbits are important too, at least in some places, together with various beetles, tubers (especially pignuts, *Conopodium*), nuts, acorns, grain and sheep and deer carrion.

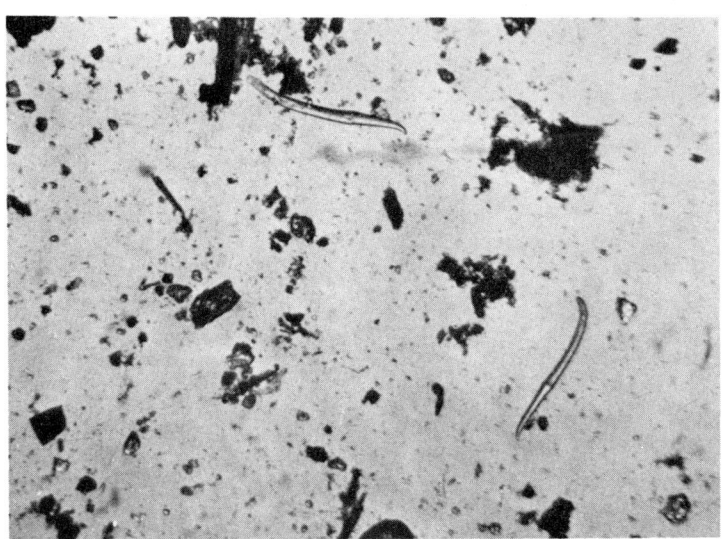

Food: remains of earthworms, the chaetae, in the faeces of a badger. Magn. ca. 20 ×.

13

Although these comparisons between the contents of badger faeces from different areas are extremely valuable, we hope also to be able to get more detailed information on the badger diet by calculating necessary correction factors, which can be applied to faecal contents in order to gain insight into the exact food composition. For this, we can use our captive animals, and also various methods to estimate the number of individual prey items that we find in the droppings. For instance, each earthworm eaten by a badger will be represented in its droppings by approximately 1,000 microscopically small bristles, the chaetae, and by one transparent gizzard-ring.

We are beginning to get an idea of how these different kinds of food are obtained from direct observations. Earthworms are usually caught on the surface, at night; hence only those

Food: dung-beetle larvae.

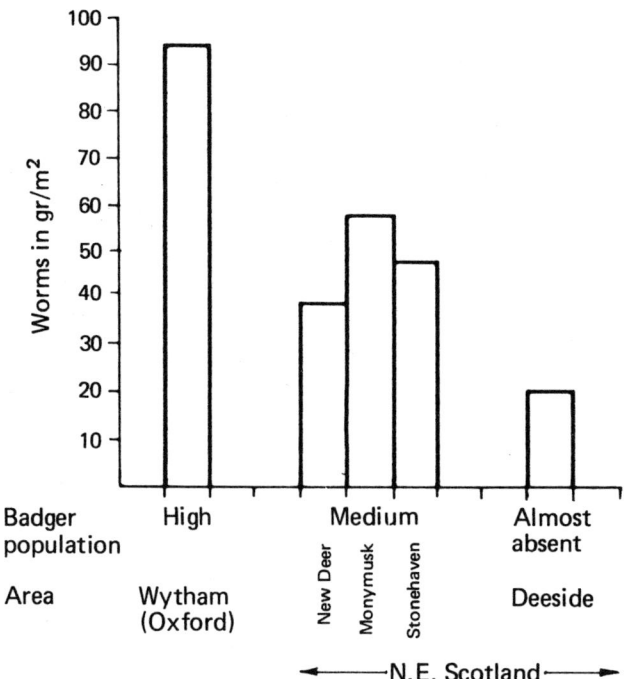

Figure 3. Biomass in (gr/m²) of earthworms in pasture in some of the study areas.

earthworm species which feed above ground are important, especially *Lumbricus terrestris*, but in Scotland also *Lumbricus rubellus*. The badgers' foraging strategy appears to be mostly adapted to catching earthworms, often feeding on pasture land where worms may occur in dense patches on the surface during humid nights. Rabbits are often taken as young ones, from the 'stops' in which they are born, but badgers appear to take a fair number of adult rabbits too, perhaps diseased animals or as carrion and it is clear that deer or sheep are only eaten as carrion. Amongst insects in the diet, dung beetles are especially important.

Whilst we are getting a good idea of what the badgers are eating in different areas and throughout the year, we are beginning to obtain estimates also of the amount of food that is available in these places. For instance, we have been looking at the presence of earthworms in pasture land in the study areas, and a pattern is beginning to emerge (Figure 3). It looks as if there is a relationship between badger numbers and earthworms, at least in some areas. A great deal more work has to be done in this field, however, and we are attempting to obtain estimates of all the other important kinds of food. A major difficulty in this is that whatever is present in an area as potential food, such as earthworms and rabbits, is not necessarily available to the badgers; earthworms are caught only under certain climatological conditions, and only part of the rabbit population can make a contribution to the badger food. However, by detailed observations of the badgers' foraging methods, we may be able to overcome these problems, at least to the extent that comparisons between areas may be possible.

Conclusions

We are trying to obtain a basic understanding of the relation between an important carnivore and its environment. We are still a long way from this target, but it is only through a study such as this, where we are looking at the animal from several different points of view (ie its ecology, its social and feeding behaviour in the field and in captivity) that we will reach our objectives. It is a long-term study, but it should be worthwhile, because of the practical necessity of understanding the ecology of carnivores, such as the badger, for the purpose of disease control and conservation and the theoretical importance of an understanding of social behaviour in the context of the animals' ecology.

Printed by Heffers Printers Ltd Cambridge England